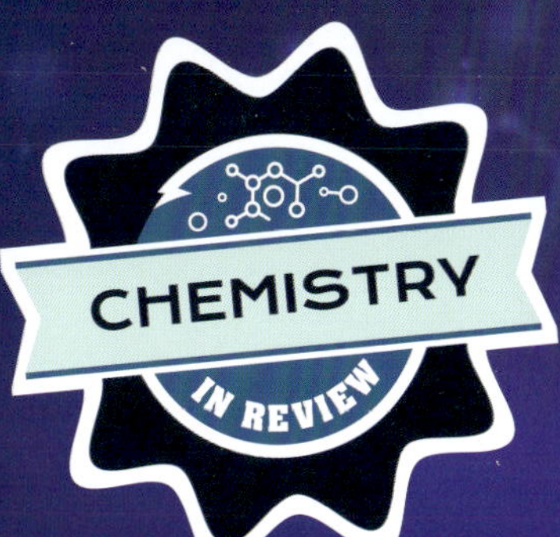

ATOMS

BY MARIE ROESSER

Please visit our website, www.enslow.com.
For a free color catalog of all our high-quality books, call toll free 1-800-398-2504 or fax 1-877-980-4454.

Library of Congress Cataloging-in-Publication Data

Names: Roesser, Marie, author.
Title: Atoms / Marie Roesser.
Description: New York : Enslow Publishing, [2026] | Series: Chemistry in review | Includes index.
Identifiers: LCCN 2024035191 (print) | LCCN 2024035192 (ebook) | ISBN 9781978542853 (library binding) | ISBN 9781978542846 (paperback) | ISBN 9781978542860 (ebook)
Subjects: LCSH: Atoms–Juvenile literature.
Classification: LCC QC173.16 .R647 2026 (print) | LCC QC173.16 (ebook) | DDC 539.7–dc23/eng/20240809
LC record available at https://lccn.loc.gov/2024035191
LC ebook record available at https://lccn.loc.gov/2024035192

Published in 2026 by
Enslow Publishing
2544 Clinton Street
Buffalo, NY 14224

Designer: Claire Zimmermann
Editor: Therese Shea

Photo credits: Cover, p. 1 (atom image) AntonKhrupinArt/Shutterstock.com; series art (molecule header image) jijomathaidesigners/Shutterstock.com; p. 5 Yavdat/Shutterstock.com; p. 9 Igor Batrakov/Shutterstock.com; p. 13 Lionel Alvergnas/Shutterstock.com; p. 19 epiximages/Shutterstock.com; p. 23 (periodic table) Peter Hermes Furian/Shutterstock.com; p. 25 (element and key) ChakSiri1969/Shutterstock.com; p. 27 Wirestock Creators/Shutterstock.com.

Printed in China

CPSIA compliance information: Batch #QSENS26: For further information contact Enslow Publishing, at 1-800-398-2504.

CONTENTS

Words in the glossary appear in **bold** the first time they are used in the text.

BUILDING MATTER

An element is a **substance** that can't be broken down into another substance. Elements are the building blocks of matter. And atoms are the **particles** that build elements! Every element is made up of its own **unique** atom.

LEARN MORE

Carbon, oxygen, and gold are some elements you might know. A single atom of an element is so small we can't see it with our eyes.

A LOOK INSIDE

An atom can't be **split** into smaller parts without losing its unique **features**. It does have parts, though. Because they're even smaller than atoms, these parts—electrons, protons, and neutrons—are called subatomic particles. They're found in certain places within atoms.

LEARN MORE

Subatomic particles are made of smaller bits too. Protons and neutrons are made up of quarks and gluons. The six kinds of quarks have names: up, down, charm, strange, top, and bottom.

At the center of every atom is a nucleus. It's made up of protons and neutrons. (Only hydrogen atoms don't have neutrons usually.) Electrons orbit, or go around, the nucleus. Their fixed paths around the nucleus are called electron shells or energy levels.

LEARN MORE

Models like this are helpful, but scientists think electrons are actually more like clouds than tiny dots.

IN CHARGE

Protons have a positive electric **charge**. Electrons have a negative electric charge. Neutrons carry no charge. Particles with opposite charges attract, or pull together. So, electrons are attracted to the positive charge of the nucleus. They're bound to the nucleus in this way.

PARTS OF AN ATOM

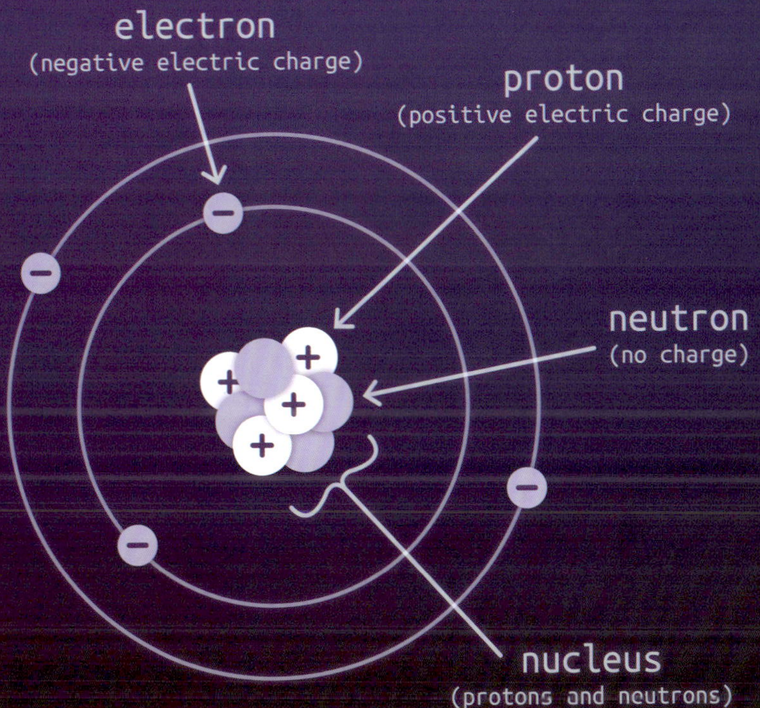

LEARN MORE

An atom with an equal number of protons and electrons has no overall electric charge.

Electrons aren't close to the nucleus in real life. They can break away from their electron shells. Sometimes they join other atoms. This changes the overall charge of the atom the electron left. It changes the charge of the atom it joins too.

LEARN MORE

If a hydrogen atom could be blown up so that its nucleus was the size of a basketball, its electron would be about 2 miles (3.2 km) away!

How does an atom lose or gain electrons? The common causes are **chemical reactions** and crashes into other atoms. An atom that loses an electron becomes positively charged. An atom that gains an electron becomes negatively charged. Charged atoms are called ions.

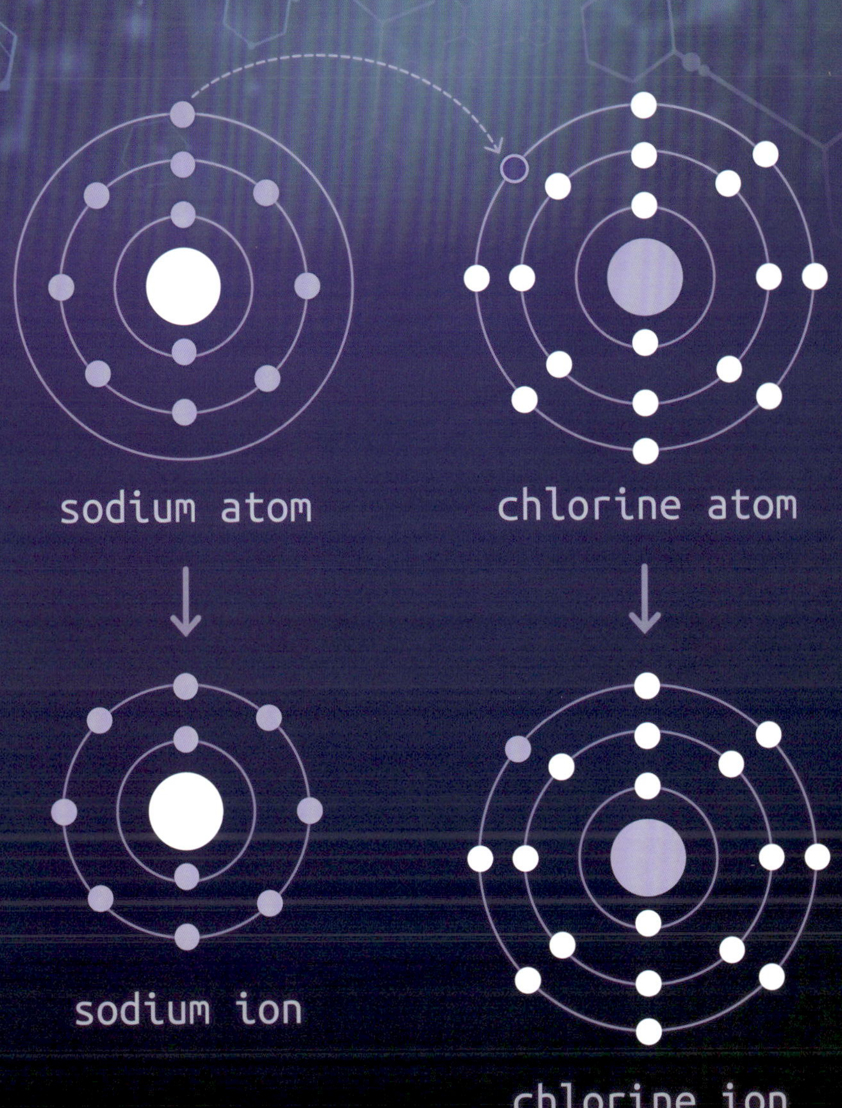

LEARN MORE

Atoms are nearly all empty space! The forces within them are strong enough to keep most particles together, however.

BUILDING MOLECULES

Atoms don't only lose or gain electrons. Sometimes, they share electrons. Electrons are found alone or in pairs in a shell. When two atoms with unpaired electrons draw close to each other, the electrons may form a pair. The atoms share these paired electrons. They form a molecule.

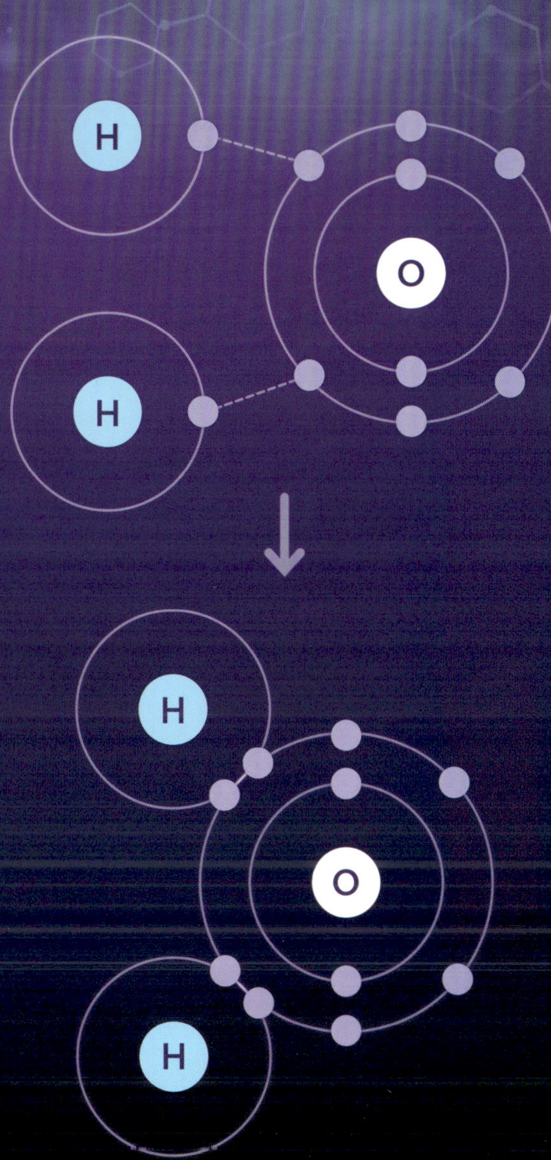

LEARN MORE

Atoms with unfilled outer shells seek a bond with other atoms. That's how this molecule of two hydrogen atoms and one oxygen atom formed. It's a water molecule!

There are 118 kinds of atoms, which make up the 118 kinds of known elements. A pure element is made up of only one type of atom. For example, molecules of hydrogen contain only hydrogen atoms. Molecules with atoms of two or more elements are called compounds.

LEARN MORE

Molecules can be made up of hundreds of thousands of atoms!

BONDING

The link between atoms in a molecule is called a chemical bond. A bond between two nonmetal atoms is a covalent bond. A bond between metals and nonmetals is an ionic bond. Metallic bonds hold atoms of metals together.

KINDS OF ATOMIC BONDS

water molecule:
covalent bond

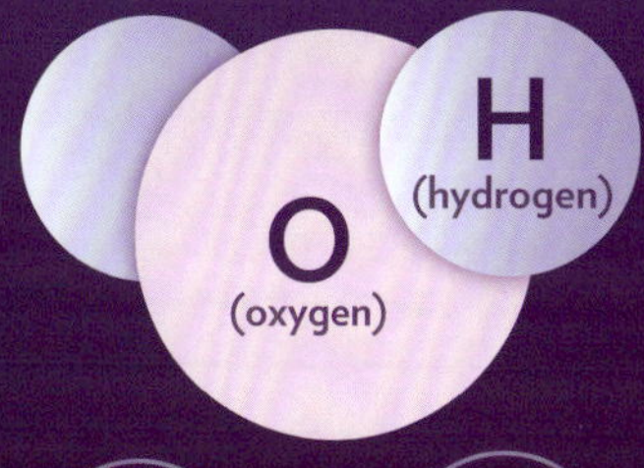

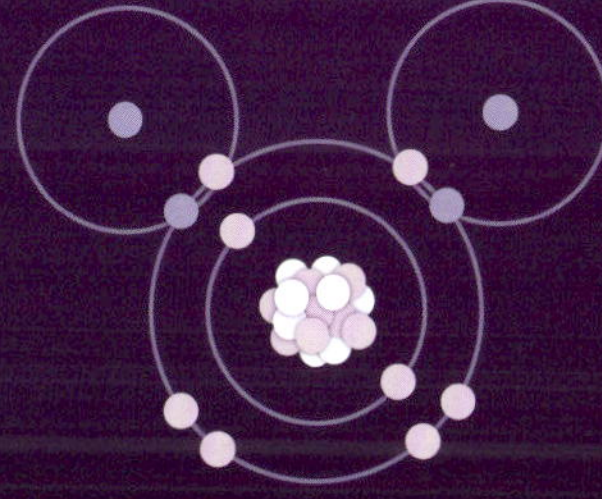

table salt:
ionic bond

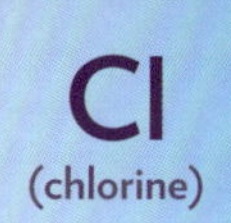

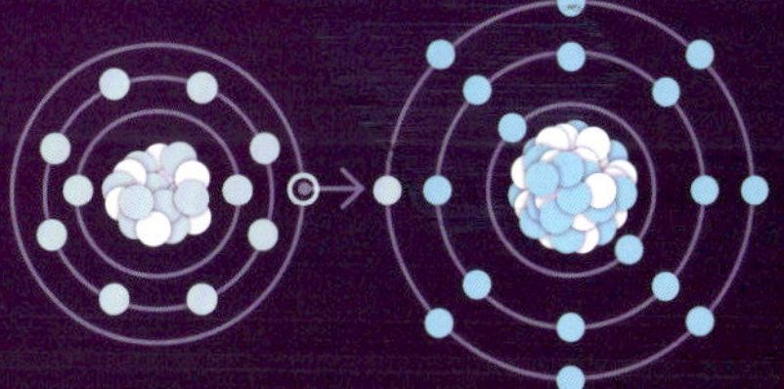

metallic bond

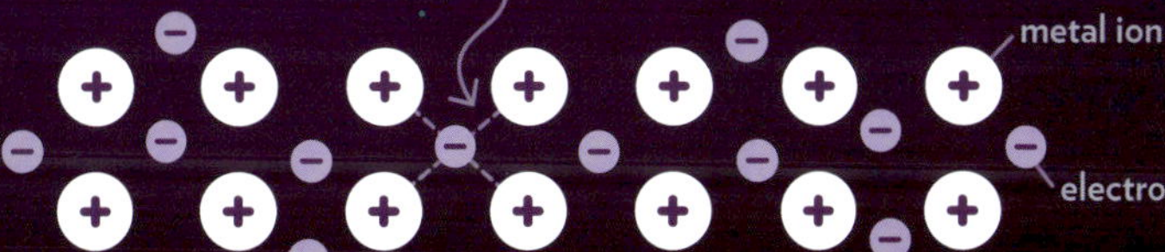

LEARN MORE

Hydrogen bonds are a weaker attraction between a hydrogen atom in one molecule and an oxygen, fluorine, or nitrogen molecule. Hydrogen bonds are important in **DNA**.

AT THE PERIODIC TABLE

The periodic table is a way of looking at all 118 discovered elements. It's organized partly by atomic number, or the number of protons in an element's atom. This proton number is different in each element, so each element has its own atomic number.

PERIODIC TABLE OF ELEMENTS

Atomic Number → 1
H ← Symbol
Name → Hydrogen

1	2	3	4	5	6	7	8	9	10	11	12	13	14	15	16	17	18
1 H Hydrogen																	2 He Helium
3 Li Lithium	4 Be Beryllium											5 B Boron	6 C Carbon	7 N Nitrogen	8 O Oxygen	9 F Fluorine	10 Ne Neon
11 Na Sodium	12 Mg Magnesium											13 Al Aluminium	14 Si Silicon	15 P Phosphorus	16 S Sulfur	17 Cl Chlorine	18 Ar Argon
19 K Potassium	20 Ca Calcium	21 Sc Scandium	22 Ti Titanium	23 V Vanadium	24 Cr Chromium	25 Mn Manganese	26 Fe Iron	27 Co Cobalt	28 Ni Nickel	29 Cu Copper	30 Zn Zinc	31 Ga Gallium	32 Ge Germanium	33 As Arsenic	34 Se Selenium	35 Br Bromine	36 Kr Krypton
37 Rb Rubidium	38 Sr Strontium	39 Y Yttrium	40 Zr Zirconium	41 Nb Niobium	42 Mo Molybdenum	43 Tc Technetium	44 Ru Ruthenium	45 Rh Rhodium	46 Pd Palladium	47 Ag Silver	48 Cd Cadmium	49 In Indium	50 Sn Tin	51 Sb Antimony	52 Te Tellurium	53 I Iodine	54 Xe Xenon
55 Cs Caesium	56 Ba Barium	57 La* Lanthanum	72 Hf Hafnium	73 Ta Tantalum	74 W Tungsten	75 Re Rhenium	76 Os Osmium	77 Ir Iridium	78 Pt Platinum	79 Au Gold	80 Hg Mercury	81 Tl Thallium	82 Pb Lead	83 Bi Bismuth	84 Po Polonium	85 At Astatine	86 Rn Radon
87 Fr Francium	88 Ra Radium	89 Ac** Actinium	104 Rf Rutherfordium	105 Db Dubnium	106 Sg Seaborgium	107 Bh Bohrium	108 Hs Hassium	109 Mt Meitnerium	110 Ds Darmstadtium	111 Rg Roentgenium	112 Cn Copernicium	113 Nh Nihonium	114 Fl Flerovium	115 Mc Moscovium	116 Lv Livermorium	117 Ts Tennessine	118 Og Oganesson

*	58 Ce Cerium	59 Pr Praseodymium	60 Nd Neodymium	61 Pm Promethium	62 Sm Samarium	63 Eu Europium	64 Gd Gadolinium	65 Tb Terbium	66 Dy Dysprosium	67 Ho Holmium	68 Er Erbium	69 Tm Thulium	70 Yb Ytterbium	71 Lu Lutetium
**	90 Th Thorium	91 Pa Protactinium	92 U Uranium	93 Np Neptunium	94 Pu Plutonium	95 Am Americium	96 Cm Curium	97 Bk Berkelium	98 Cf Californium	99 Es Einsteinium	100 Fm Fermium	101 Md Mendelevium	102 No Nobelium	103 Lr Lawrencium

LEARN MORE

Scientists have created new elements by adding a proton or neutron to another element. It doesn't always work and often doesn't last long when it does.

Periodic tables provide **information** about each element. They often give the symbol for each element, which is a letter or letters that stand for the element's name. Atomic mass is the number of protons and neutrons in an atom. It's measured in atomic mass units.

PERIODIC TABLE KEY

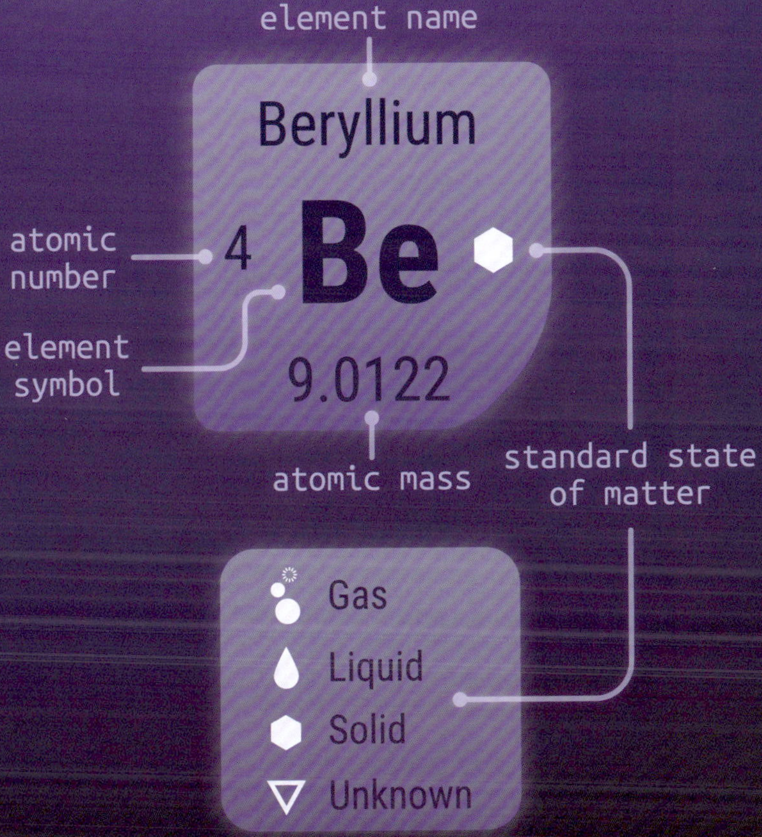

element name

Beryllium

atomic number — 4

element symbol — Be

9.0122

atomic mass

standard state of matter

- Gas
- Liquid
- Solid
- Unknown

LEARN MORE

The color of the element on the table also gives information about the element. Periodic tables can be different, so always look at the key.

NUCLEAR FISSION

Scientists have figured out how to split the nucleus of some atoms. When the nucleus of an atom breaks into two or more parts, nuclear fission takes place. It can produce a great amount of energy, or power. Nuclear fission in power plants is used to create electricity.

LEARN MORE

People have also used nuclear fission to make powerful and deadly nuclear **weapons**.

NUCLEAR FUSION

Nuclear fusion happens when two atoms combine to form a new atom. This, too, produces great amounts of energy. Nuclear fusion happens naturally in stars. That's how our sun gives off light and heat energy, making life possible on Earth!

NUCLEAR REACTIONS

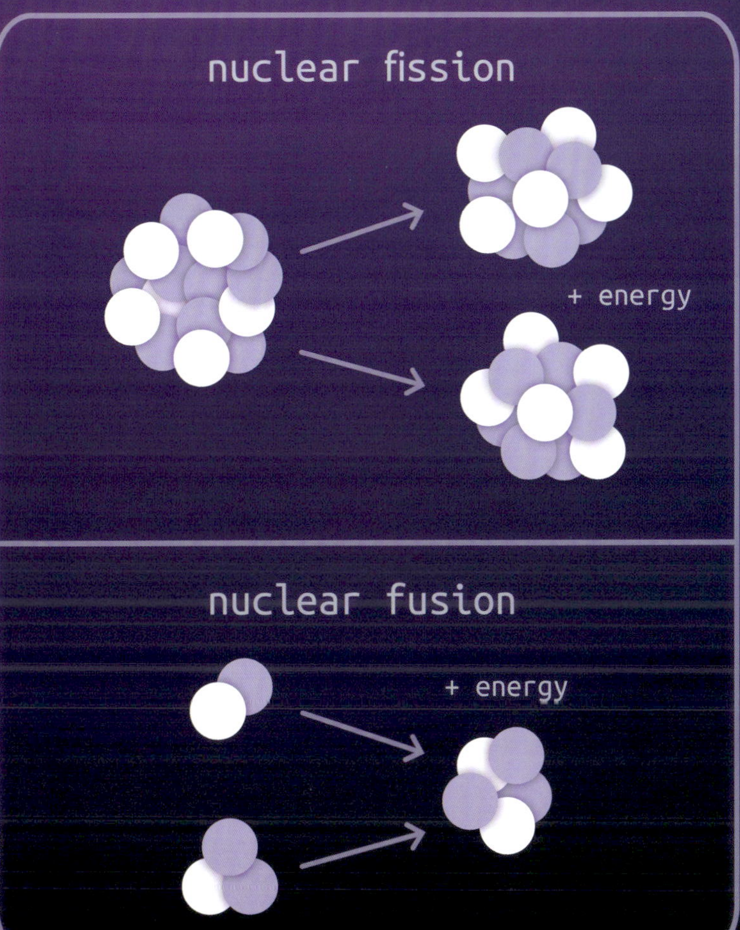

LEARN MORE

Scientists are hoping nuclear fusion may provide for people's energy needs in the future.

AN ATOMIC VIEW

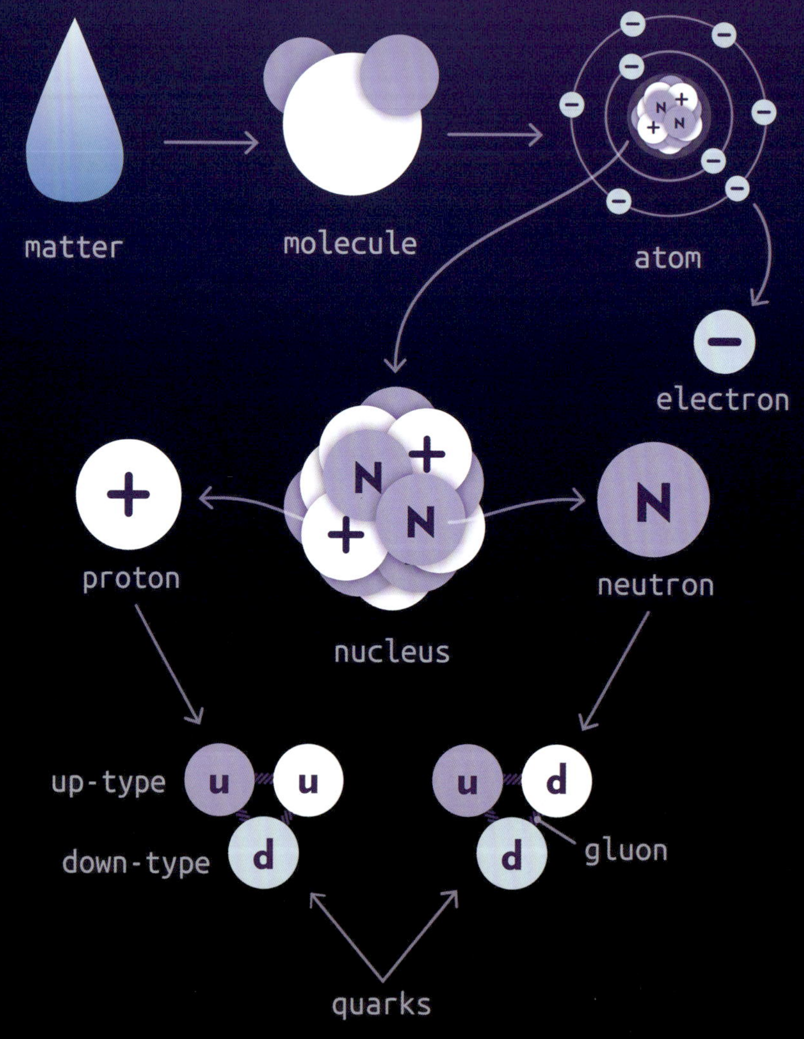

GLOSSARY

charge: An amount of electricity.

chemical reaction: A change that occurs when two or more kinds of matter combine to form a new kind.

DNA: Part of the body that carries genetic information, which gives the instructions for life.

feature: An interesting or important part, look, or way of being.

information: Knowledge obtained from study or observation.

particle: A very small piece of something.

split: To break into parts, especially in halves.

substance: A certain kind of matter.

unique: One of a kind.

weapon: Something used to fight an enemy or to cause someone or something injury or death.

FOR MORE INFORMATION

BOOKS

Rusick, Jessica. *Investigating Atoms & Molecules.* Minneapolis, MN: Checkerboard Library, 2023.

McKenzie, Precious. *The Micro World of Atoms and Molecules.* North Mankato, MN: Capstone Press, 2022.

WEBSITE

Atoms and Their Sizes
www.amnh.org/exhibitions/permanent/scales-of-the-universe/atoms
Learn about atomic size and about certain kinds of atoms on the American Museum of Natural History's site.

INDEX